RAND

Suggestions for Strategic Planning for the Office of Nonproliferation Research and Engineering

James L. Bonomo

Prepared for the
Office of Nonproliferation Research and Engineering

National Defense Research Institute

The research described in this report was sponsored by the Office of Nonproliferation Research and Engineering of the National Nuclear Security Administration, Department of Energy. The research was conducted in RAND's National Defense Research Institute, a federally funded research and development center supported by the Office of the Secretary of Defense, the Joint Staff, the unified commands, and the defense agencies under Contract DASW01-01-C-0004.

Library of Congress Cataloging-in-Publication Data

Bonomo, James.
Suggestions for strategic planning for the Office of Nonproliferation Research and Engineering / James Bonomo.
p. cm.
"MR-1499."
Includes bibliographical references.
ISBN 0-8330-3142-2
1. Nuclear engineering—United States—Planning. 2. United States. National Nuclear Security Administration. Office of Nonproliferation Research and Engineering—Planning. 3. Nuclear nonproliferation—Research—United States—Planning. I.Title.

TK9023 .B65 2002
621.48'072073—dc21

2002021306

Published 2002 by RAND
1700 Main Street, P.O. Box 2138, Santa Monica, CA 90407-2138
1200 South Hayes Street, Arlington, VA 22202-5050
201 North Craig Street, Suite 202, Pittsburgh, PA 15213-1516
RAND URL: http://www.rand.org/
To order RAND documents or to obtain additional information, contact Distribution Services: Telephone: (310) 451-7002; Fax: (310) 451-6915; Email: order@rand.org

Preface

The Office of Nonproliferation Research and Engineering, part of the recently formed National Nuclear Security Administration within the Department of Energy, conducts wide-ranging research for diverse end users at the federal and local level. As such, the Office faces a number of challenges in choosing how to best serve its users' various needs while justifying its budget.

To help meet these challenges, this report suggests specific changes in the Office's planning of its research program. These suggestions are based on an understanding of the Office's activities and a survey of available planning methods. This report grew out of support provided to an advisory committee for the Office of Nonproliferation and National Security, an organization that includes the Office of Nonproliferation Research and Engineering. This report describes alternative methods for planning the research activities of the Office of Nonproliferation Research and Engineering and suggests which of those methods seem particularly appropriate to resolving the specific challenges the Office faces. As such, this report should be of interest not only to those within the Office of Nonproliferation Research and Engineering but also to anyone responsible for planning similar research programs.

The research for this report was conducted for the Office of Nonproliferation Research and Engineering of the National Nuclear Security Administration, Department of Energy, within the International Security and Defense Policy Center of RAND's National Defense Research Institute, a federally funded research and development center sponsored by the Office of the Secretary of Defense, the Joint Staff, the unified commands, and the defense agencies.

Contents

Table

Summary

The Office of Nonproliferation Research and Engineering of the National Nuclear Security Administration has a large and diverse research program. The planning of research activities in this program presents a number of unusual challenges, which prompted this examination of the Office's planning methods.

Unlike many other research programs, the Office of Nonproliferation Research and Engineering[1] program has numerous disparate users of its products. Its users range from other offices within the National Nuclear Security Administration to offices within other federal agencies, such as the Department of Defense and Department of Justice, to "first responders," such as fire departments, police departments, and medical personnel at the scene of an emergency.

The activities conducted within the research program are important to the users' missions, which range from verifying arms control agreements to coping with domestic terrorist acts involving biological or chemical weapons. In the case of many users, the Office of Nonproliferation Research and Engineering is the only available source of research and development that is focused on their needs.

Additionally, the Office's research program includes an unusually wide range of activities. Some of its work involves basic research in fundamental science, such as molecular biology or material science. At the other end of its spectrum of activities, the program produces flight hardware for satellite systems.

The challenges the Office faces are compounded by its being part of an institution in a state of flux. The Office of Nonproliferation Research and Engineering is now part of the new, semiautonomous National Nuclear Security Administration.[2] As with any new organization, the National Nuclear Security Administration has

---

[1]The Office's charter is to "provide leadership in policy support and technology development for international arms control and nonproliferation efforts" (Department of Energy, 2000b).

[2]The National Nuclear Security Administration officially began operations on March 1, 2000. As stated on the National Nuclear Security Administration Web site (http://www.nnsa.doe.gov/about_nn.asp), the Administration's mission is to "(1) Enhance United States national security through the military application of nuclear energy. (2) Maintain and enhance the safety, reliability, and performance of the United States nuclear weapons stockpile, including the ability to design, produce, and test, in order to meet national security requirements. (3) Provide the United States Navy with safe, militarily effective nuclear propulsion plants and to ensure the safe and reliable operation of those plants. (4) Promote international nuclear safety and nonproliferation. (5) Reduce global danger from weapons of mass destruction. (6) Support United States leadership in science and technology."

had to establish its own procedures, which must be matched to its mission and goals.

All of these challenges present two related problems for the Office's research program: First, program planners must fully understand the needs of the program's diverse customers. Second, program planners must not only choose which of those needs the program will address, but also defend their choices convincingly during the annual budgetary process. In addition to the leadership of the National Nuclear Security Administration, first the leadership of the Department of Energy, then the U.S. Office of Management and Budget, and finally the Congress all must approve any proposed program research.

These agencies and Congress also place other demands on the Office of Nonproliferation Research and Engineering. The Government Performance and Results Act of 1993 mandates that all agencies prepare an overall strategic plan, as well as annual performance plans and program performance reports. The latter are evaluations of how well the goals of the performance plan were met. Additionally, annual budgetary instructions from the Office of Management and Budget direct federal agencies to concentrate on certain aspects of their research programs. Other, less-formal demands are placed on agencies, including the Office of Nonproliferation Research and Engineering, by reports from nationally chartered review groups, such as the National Research Council and the National Science Board.

Some planning methods appear to be particularly promising for helping the Office of Nonproliferation Research and Engineering resolve its challenges and problems. These methods were identified through a review of planning methods for research and development that are used in other federal agencies and private industry and through a review of methods described in the literature. By using two recommended sets of methods, which are discussed in Chapter 3, both of the major problems the Office faces could be addressed, at least in part, without diverting a large amount of funding or managerial attention.

The first set of planning methods would be focused on strengthening the Office's connection with the various end users of its research. A stronger connection would both help the research program better understand the needs of its users and also help the users accept the products of the research more readily. To forge this connection, users could be involved in developing the concepts that motivate the Office's research; a method called "Concept Option Groups"[3] (discussed in

[3]A Concept Option Group is a small group that includes both selected users and selected developers or researchers. The group's function is to develop new concepts for accomplishing the users' tasks by drawing on potential technical developments that appear plausible to the developers.

Chapter 2) is well suited to that role. Alternatively, users could be involved in assessing the relevance of certain ongoing research. Both of these roles appear to be applicable to various parts of the overall program.

The second set of planning methods would be focused on both demonstrating and increasing the quality of the program through expansion and systematization of the peer reviews[4] already used in parts of the program. When appropriate, these peer reviews could also be combined with the users' relevance assessments from the first set of planning methods.

Together, these methods would offer five distinct advantages: (1) They are largely consistent with existing methods used within the Department of Energy and the Office of Nonproliferation Research and Engineering. Thus, they should face no significant bureaucratic obstacles. (2) The methods would increase the visibility of the diverse users to the leadership of the National Nuclear Security Administration. This should aid the leadership's understanding of the program. (3) The methods match important trends across the federal government, which include an increased role for peer review. (4) The methods should increase the utility of the research by more closely coupling it to actual needs. (5) The methods also should produce additional information on the projects within the program. The information could be used by the program's leadership in making programmatic choices that inevitably arise, such as when budgets are changed by an administration or by Congress on short notice.

In considering all the various planning methods that have been studied, those that appear to be both relatively inexpensive and that offer the most potential to ameliorate the challenges faced by the Office of Nonproliferation Research and Engineering are (1) the addition of selected Concept Option Groups and relevance assessments and (2) the expansion of the peer review process. This report examines those options.

The group tightly focuses on a single task, iterating concepts and technical possibilities, to eventually produce plans for specific technical programs that, if successful, would enable the new concepts.

[4]*Peer review* in this context would primarily be used to judge the technical merit of the research, with perhaps a secondary focus on management or institutional capability to execute the plan. The peers would be researchers in the same (or a closely related) field who were not active participants in the research program. Such peer review provides external validation of the quality of the work, which can be used not only for setting internal priorities, but also for defending the program to higher officials in the National Nuclear Security Administration or to Congress.

Acknowledgments

This report would not have existed without the support and encouragement of Dr. Robert Nurick, who led the project at RAND. Likewise, it could not have been written without the interested involvement of Robert Waldron, the Assistant Deputy Administrator for the Office of Nonproliferation Research and Engineering, and Gerald Kiernan, his deputy. I would also like to thank Dr. Robert Hirsch for his careful review. Of course, any errors that remain are the sole responsibility of the author.

1. The Challenges of the Office of Nonproliferation Research and Engineering

The Office of Nonproliferation Research and Engineering within the National Nuclear Security Administration (NNSA) of the Department of Energy (DOE) has a wide array of responsibilities and activities. Its responsibilities range from supporting specific technical provisions of formal treaties to providing technologies for use by domestic first-responders to emergencies, such as police and fire department personnel, who may face terrorist use of nuclear, chemical, or biological materials. The Office's activities range from basic research in genetics and protein synthesis to the production of operational satellite subsystems.

Adapting to an Institution in Flux

The Office of Nonproliferation Research and Engineering must fulfill its responsibilities within a dynamic and complex institutional context. Since the creation of the National Nuclear Security Administration in March 2001, the Office has been part of that newly formed institution. The new, semi-autonomous administration within the Department of Energy is instituting its own planning, programming, budgeting, and evaluation system, which will likely differ from the existing DOE system (U.S. Department of Energy, 2000a).

At the same time, the Office remains a part of the DOE, and its activities presumably will still be included in one of the DOE's four research and development (R&D) portfolios (U.S. Department of Energy, 2000b).[1] Additionally, the Office will continue to tap scientific expertise within the non-NNSA parts of the DOE (U.S. Department of Energy, 2000c). In particular, existing research programs at non-NNSA laboratories, such as the Pacific Northwest National Laboratory, are very likely to continue to be funded by the Office. These factors will likely demand that some changes be made in management as the NNSA becomes more fully developed, so that the NNSA has the authority over the programs that Congress has mandated.

---

[1] I discuss the concept of R&D portfolios in the section "Building on the Process" later in this chapter.

The Office will also be included in strategic and performance plans, as mandated by the Government Performance and Results Act of 1993, and in related annual evaluations as part of both the Department of Energy and NNSA (National Science Board, 2000). The planning and evaluation processes may each demand a different aggregation or interpretation of the activities of the Office. Additionally, the processes may emphasize different objectives and thus different responsibilities for the Office.

A Wide Range of Users and Applications

The responsibilities of the Office of Nonproliferation Research and Engineering constitute national objectives that the Office rarely accomplishes alone. Rather, the role of the Office is to provide systems, technologies, and/or knowledge to a wide range of end users. In a few cases, the users are within the DOE or NNSA, such as when the Office develops measurement equipment or methods for safeguarding fissile material in the United States or Russia. More often, the users are outside the Department of Energy in the national security community, in federal law enforcement agencies, and even in local agencies, such as fire departments or transit systems.

The Office also supplies technical expertise for negotiations on existing or potential international agreements. Some of these agreements involve limitations on fissile material for which the obvious expertise of the nuclear weapons laboratories (which are under NNSA oversight) is applied; but other international agreements involve areas, such as an inspection protocol for the Biological Weapons Convention,[2] that tap other expertise within the Department of Energy labs. However, classification requirements limit the open discussion of some of the Office's other activities and users.

This wide range of diverse users with different missions presents unusual problems for the Office, which must correctly interpret the needs of these users as it develops new technologies. These challenges become more difficult as users' needs shift and change with time, such as when negotiations are proceeding, when an operational problem becomes better defined, or simply when the leadership and perspective of some groups change. For the Office's many end users who are outside of the NNSA, these constant changes demand cross-institutional connections that the National Aeronautics and Space Administration centers or industrial laboratories, for example, rarely experience.

[2]As a point of interest, the Biological Weapons Convention has been in force since March 26, 1975. States can become participants at will. More than 100 nations are now party to the treaty (see http://www.state.gov/www/global/arms/treaties/bwc1.html#3).

A Wide Variety of Activities

Some activities within the Office of Nonproliferation Research and Engineering can be characterized as "early-applied" (or "targeted-basic") research. This stage of research is characterized by a real, but still quite loose, coupling to an eventual application. Some of this early-applied research is relatively small in scale and is performed within both the laboratories of the Department of Energy and in research universities. Examples of this sort of research can be found in most program areas of the Office; they range from seismic analysis and data gathering to detector development and flow simulations to basic biological studies of pathogenic organisms. Some of the early-applied research is larger in scale and therefore is more likely to be performed within national laboratories or private-industry facilities rather than in academic settings. Examples of larger-scale, but still rather basic, research would include the development of ultra-low-level light detectors or large-scale fluid flow simulations.

Other activities within the Office can be characterized as large-scale demonstrations, most often centered at a national laboratory, which assumes the lead for the demonstration. A particularly striking example of a large-scale demonstration is the Multispectral Thermal Imager (Office of Deputy Administrator for Defense Nuclear Nonproliferation, 2001). This effort will encompass every stage of research and development from initial modeling to the fabrication and flight of an imaging satellite with 15 spectral bands ranging in wavelength from 0.45 micrometer to 10.70 micrometer. Accurate radiometric calibration of this satellite system should enable qualitatively new observations of natural and man-made phenomena, which will have applications ranging from nonproliferation of chemical, biological, or nuclear weapons to crop monitoring, for example. Even smaller-scale efforts, such as CALIOPE (Chemical Analysis by Laser Interrogation of Proliferation Effluents), were led by a laboratory or group of laboratories because of the range of activities that were involved in the project (Skiby, 1998).

The Office of Nonproliferation Research and Engineering also produces operational hardware, an example of which is the satellite sensors for detecting nuclear detonations in the atmosphere or in outer space. The Office has a long history of supplying such sensors, which has allowed it to support the full range of activities on the development of related technologies, from applied research and demonstrations to the production of sensor subsystems for integration into operational satellites.

Building on the Current Process

The Department of Energy recently embraced the idea of "R&D portfolios" as an organizing principle across the department. The DOE defines an R&D portfolio as ". . . a comprehensive Description of all of the Department's research activities into one coherent strategic framework . . ." which includes ". . . the goals, strategies for accomplishing these goals, and the highest priority R&D needs supporting these strategies" (U.S. Department of Energy, 2000b). This concept is an ideal fit for the research program of the Office of Nonproliferation Research and Engineering. In fact, the Office may be able to use this concept even more broadly than can the DOE as a whole.

The Office actually has a "portfolio of portfolios." An ordinary R&D portfolio, such as one fitting the DOE's definition, encompasses a variety of goals and, typically, covers the varying periods of time between initiating some work and providing final finished results to a user. In addition, the Office has an unusually wide range of customers and types of programs. For each of these customers and programs, the Office must create a portfolio of research activities, creating in aggregate a portfolios of portfolios.

Ideally, the Office of Nonproliferation Research and Engineering should be able to use the taxonomy of methods already established by the Department of Energy to articulate its research program. The aim of every program within the Office relates to a single goal in the Department of Energy Strategic Plan: "Reduce the global danger from the proliferation of weapons of mass destruction" (U. S. Department of Energy, 2000d).

Two related performance measures, also from the Department of Energy Strategic Plan and developed in part to satisfy provisions of the Government Performance and Results Act, are relevant to the Office:

- Provide leadership and technical support to interagency nonproliferation and arms control efforts to strengthen the international nonproliferation regime.
- Demonstrate technologies to detect proliferation of weapons of mass destruction (U.S. Department of Energy, 2000d).

These performance measures are further elaborated in the Department of Energy Strategic Plan to include support for the implementation of the Comprehensive Test Ban Treaty and support for a range of planned demonstrations, such as "performing an airborne demonstration of new technology for detecting WMD [weapons of mass destruction] proliferation by 2005" (U.S. Department of

Energy, 2000d, p. 50). The existing Department of Energy Strategic Plan describes strategies for achieving its goals in keeping with the Government Performance and Results Act. Many of these strategies, such as the following, directly support the Office's research program:

- Conduct analyses and technology-development efforts for transparency activities (focusing on verifying warhead dismantlement to help ensure that nuclear reductions between the United States and Russia are transparent to both sides and irreversible).
- Develop and demonstrate technologies that are needed to remotely detect the early stages of a proliferant nation's nuclear weapons program.
- Improve capabilities to locate, identify, and characterize nuclear explosions.
- Produce operational satellite-based sensor systems for monitoring nuclear explosions.
- Improve the U.S.'s capability to detect the proliferation of chemical and biological agents at an early stage and minimize the consequences from the use of chemical or biological agents (U.S. Department of Energy, 2000d, p. 51–52).

Clearly, the programs of the Office of Nonproliferation Research and Engineering are well supported by the Department of Energy Strategic Plan. This support partly reflects some recognition within the DOE of the Office's goals. Building upon these existing terms within the plan should be the easiest way to allow individuals outside of the DOE to understand the variety of national goals supported by the Office's research program.

Two elements that are important to the Office's planning are currently missing from the Department of Energy Strategic Plan. One missing element is an explicit recognition of the full range of users or customers and their specific objectives for each strategy. Recognizing the existence of the customers who are outside of the Department of Energy is important in understanding the extent of the demands on the Office. The other missing element is the range of activities within the program of the Office, from fundamental science to production. This wide range of activities not only reflects a real variation in the ability of current technology to support solutions to certain problems, but also reflects a balance being struck between modest near-term solutions and riskier farther-term but potentially more robust solutions to other problems.

These three factors—recognizing the broad range of users, understanding the wide range of research activities, and using the existing strategies of the

Department of Energy—are best accommodated using the concept of a "portfolio of portfolios" described earlier. The Office of Nonproliferation Research and Engineering can use that concept to justifiably explain that its research program must strike a balance across not only the time horizon or measure of risk that is usually implicit with the term "portfolio," but also across fundamentally different demands, thereby satisfying the needs of different users. Because the users and their needs are unusually diverse, there is an unusual amount of diversity within the program, as compared with the research programs of other federal agencies. This diversity is an important key to understanding the Office's program, which is why the idea of a portfolio of portfolios is so useful.

Naturally, a fuller explanation of the diverse demands on the program requires more-sophisticated analytic tools. Equally, the diverse portfolio of portfolios requires a range of other tools for its planning and management beyond those in the Department of Energy Strategic Plan. Chapter 2 addresses these tools.

2. Alternatives for Strategic Planning

The rich literature on strategic planning reflects the importance of having an effective plan, the large effort needed to create a plan, and the range of planning alternatives embraced by different organizations. Some planning methods are based on a long governmental history (for example, the Planning, Programming, and Budgeting System of the Department of Defense). Most planning methods used in and out of government draw on the large number of strategic planning applications in the business world. While the highly competitive business market for planning methods produces genuine innovations in methods, it also produces mere fads. In this chapter, I focus on strategic planning methods that I feel have enduring value.

A Taxonomy for Strategic Planning Methods

With such a wide range of strategic planning methods in use, it becomes important to place the methods described here in a broader context. Table 2.1 categorizes and briefly describes a range of strategic planning methods. The table is adapted from a report done for the U.S. Air Force by Davis and Khalilzad (1996). In Davis and Khalilzad, the authors examined a wide variety of methods that might be used to guide planning processes within the U.S. Air Force. The authors clearly illustrate the range of subjects for analysis and corresponding planning methods that could be called "strategic planning."

The National Nuclear Security Administration may well end up conducting planning activities at all the levels shown in Table 2.1. The Office of Nonproliferation Research and Engineering may in turn be required support such activities as they evolve, or at least be aware of them and their implications for the program of the Office.[1]

For its own use, the Office of Nonproliferation Research and Engineering would have the greatest interest in the subjects and methods listed in the third through fifth rows in the table. The subjects of the first two rows of the table would be addressed by the national leadership of the United States and, in more detail, by

[1]For a more complete description of all of these methods, see Davis and Khalilzad (1996). Six other references have proved particularly useful in this research: Kent and Simons (1994), Dewar et al. (1993), Mintzberg (1994), Schwartz (1991), and Wack (1985a and 1985b).

Table 2.1

Range of Planning Methods

Subject for Analysis	Methods That Are Useful in Analysis	Comments
National Security Strategy (NSS) and National Military Strategy (NMS)	Uncertainty-sensitive planning (USP) (including examinations of alternative strategic environments, budget levels, and priorities among national interests) Alternative futures	Premium is on open-minded and divergent thinking, followed by synthesis of an integrated, high-level strategy. Output from creative and analytic exercises may or may not produce clear-cut decisions but will include insights affecting plan-level decisions. Focus is on developing alternative visions of the future, for both the external environment and the national mood (e.g., neo-isolationism).
Joint Missions and Operational Objectives	Objective-based planning (going from strategies-to-tasks) conducted for a wide range of circumstances	Premium is on top-down structured analysis. Output is a taxonomy of well-defined tasks to be accomplished, which are motivated by the national strategy and its priorities.
Joint Tasks	Objective-based planning	Premium is on translating abstract functions into concrete tasks suitable for practical management.
Operational Concepts	Concept Option Groups Comparative systems analysis Organizational-viability analysis	Premium is on creative but pragmatic work producing concrete system concepts for accomplishing the various tasks and missions, followed by objective trade-off analyses to choose among competitive concepts, while taking into account organizational willingness to first seriously test a concept and then, if it is successful, implement it.
Individual Programs	Program analysis (especially marginal analysis)	Objective is to translate operational concepts into programs for procurement, doctrinal change, training, and so forth, and to analyze programs.
National Security Program and Posture	Exploratory modeling Adaptive planning (also known as capabilities-based planning), which includes scenario-space contingency analysis and strategic portfolio analysis Stressful scenario sets Assumption-based planning	Objective is to assess programs and postures, both for the full range of operations and for different budget levels, against a broad range of future challenges (so-called "scenario-space analysis") and against needs to influence the strategic environment. These methods can also simplify the expression of requirements for management of programs and other activities and allow the leadership to review and amend plans to better cope with uncertainty.
Long-Range Programs	Objective-based planning (variant of the items in the second and third rows of this table) Exploratory modeling Assumption-based planning Long-term program and budget analysis	Special features include emphasis on next-generation capabilities whose applications are not fully understood. Pursuing the enabling technologies is an important step in setting the stage for the creation of a new military, often referred to as the "military-after-next." Compared with other applications of these general methodologies, there is less emphasis on the use of scenarios.

the departmental leadership and the administrator of the NNSA. These first two rows emphasize the large-scale balancing of missions and objectives that, once set, should guide the Office's more-detailed strategic planning. The bottom two rows in the table address large-scale balancing of missions and objectives—the choices, priorities, and assignments to agencies—which occur over longer time periods, with potentially large consequences for the overall objectives for an organization.

Applying the Taxonomy to the Office of Nonproliferation Research and Engineering

After its broad national and departmental guidelines and strategy are in place, planning becomes an important function of an organization such as the Office of Nonproliferation Research and Engineering. As Table 2.1 shows, the primary tools such an organization uses are Objective-Based Planning; Concept Option Groups, which is a method related to objectives-based planning for defining new system concepts (see Kent and Simons, 1991); trade-off analyses for those new concepts; and program analysis methods. Although these tools have a narrower scope than the methods listed in Table 2.1, they nevertheless are highly diverse.

Objective-Based Planning

Objective-based planning methods are important to the Office of Nonproliferation Research and Engineering for several reasons. First, as noted in the table, these methods explicitly connect the program activities with national objectives. In a sense, the Department of Energy Strategic Plan attempted to do this, albeit without all the linkages to national objectives involving the Office. A more complete articulation of the linkages connecting the disparate elements of the Office's research program to national objectives could be useful in and of itself if only to explain the purposes of the program to anyone unfamiliar with it. In particular, a full, formal articulation of the connection would be useful to the Office in explaining its work internally to various researchers and externally to any new leadership in the DOE or in Congress.

In fact, a recent review of the Office's research program starts with a list of national goals and the connection between those goals and the research program (U.S. Department of Energy Nonproliferation and National Security Advisory Committee, 2000, pp. 1–4). A discussion of that connection clearly is essential to any examination of the Office's research program and to input on any planning of the program. The exact method that is used is much less important than the creation and articulation of the linkages between national goals and program research.

One method of objective-based planning, which was developed at RAND, is called "strategy-to-tasks" (see Thaler, 1993, and Kent, 1989). The name of the method simply refers to the fashioning of logical connections from national strategy, as formally enunciated in various guidance documents and public pronouncements, to the specific tasks an organization does or is prepared to do—in this case, the research tasks undertaken by the Office. RAND had begun to explicitly fashion such a strategy-to-tasks analysis of the Office's research program before being redirected to higher-priority needs. That partial effort did nevertheless demonstrate that this method could be applied to the Office's diverse research program.

In principle, methods such as strategy-to-tasks can be applied at different levels of detail, depending on the purpose for applying the method and the audience for the resulting connection from strategy to tasks. At one extreme, every project that the Office supports could be linked to some national objective. Demanding that each project make that connection can be useful in focusing the attention of individual researchers and managers on the real objectives rather than primarily on technical developments. Alternatively, applying the method to an aggregation of projects that is collectively and often informally called a "program" should suffice for an external audience seeking a quick overview of the Office's nonproliferation program. The required resolution for planning should be determined by top Office management, depending upon the role that the managers foresee the process playing.

The strategy-to-tasks method can also be applied internally to the actual planning of research. For instance, the method has been used at RAND as a framework to help cross-functional teams identify new research needs. The focus of these teams is on developing new concepts for accomplishing an operational task; thus, this method has been given the name "Concept Option Groups" (COGs) (Kent and Simons, 1991). To identify operational concepts, such teams must include the appropriate users from the particular communities or groups responsible for accomplishing the task. Because the new concepts often depend upon having some additional ability, the teams must also include technical experts who can suggest potential new systems that could provide the missing key ability that would make the operational concept feasible. The technical experts are also essential for another role—they must identify the technical developments needed to create the systems. Often, a small support group is also used to perform simple trade-off analyses among the concepts. With its emphasis on operational concepts, this method obviously is focused on the late stages of technical development, from proof-of-principle experiments to prototype demonstrations. Fortunately, these later stages of technical development overlap much of the activity supported by the Office. The teams of users and technical experts also could be tasked to identify new initiatives for the Office in areas selected by the Office leadership—perhaps to identify new, large-scale demonstrations within the chemical and biological defense program or a new operational focus within the radiation detection program.

The benefits of using the teams in these ways are twofold. First, there is the potential of uncovering genuinely new ideas through the direct interaction of users and developers. The formation of a cross-functional team can help to avoid the misunderstandings that tend to arise with a more-distant, formal process. Second, the use of such teams can also increase support for technical initiatives within the user community. Sometimes that support is manifested in partial funding of the development or support for field tests of new systems. Even if the user community is unable to financially support the technical research, as would be the case with some local governmental users supported by the Office, user-community involvement in research planning can encourage users' acceptance of any systems that eventually succeed. That is, a successful COG can avoid the problem of researchers springing a solution on a user community without considering the reaction of the users or without even fully understanding their needs and constraints. The use of cross-functional teams seems to be especially important to a research program such as the Office's with its broad and diverse group of potential users.

Methods for Research and Development Management

A subset of the general literature on management describes methods of managing technical enterprises (Beattie and Reader, 1971; *R&D Productivity . . .*, 1974; Camp, 1989; and Steele, 1989). This subset usually emphasizes motivating research workers and maintaining a healthy institution. To some degree, these are the Office's concerns. A Nonproliferation and National Security Advisory Committee report discusses the need for the Office to support and to even be the steward for what the committee called the "NN (nonproliferation and national security) Tech Base" (Department of Energy Nonproliferation and National Security Advisory Committee, 2000, p. 25f). The report places an emphasis on the support of a few specific activities related to these concerns, notably the availability of funding for advanced concepts research. But neither that report nor other reviews have called for the Office to be involved in the details of maintaining a motivated and skilled work force.

Such details, while they are important, are naturally the concern of the research institutions themselves. For most of the efforts within this research program, those institutions are the various national laboratories of the NNSA and the Department of Energy. The Office of Nonproliferation Research and Engineering does not have the manpower, information, or authority to perform such functions. Consequently, this report does not address most methods for managing technical enterprise.

A method of particular interest is the concept of a "technology road map." Technology road maps appear in the planning literature as a means of communicating the goals and development schedule of a research program in terms that are meaningful to both the individual researchers, who understand the technical details, and the technical managers, who are trying to orchestrate some important advance for their organization. As such, the road maps should logically derive from a framework, such as the output from a COG discussed earlier. Certainly, the level of detail in a road map should reflect the road map's eventual use. In most cases, however, a technology road map should be predicated on a precise connection between specific operational concepts and the technical accomplishments that are necessary before those concepts become practical.

The current Department of Energy Strategic Plan highlights the usefulness of technology road maps (U.S. Department of Energy, 2000d). Although future strategic plans may change as the National Nuclear Security Administration becomes more firmly established, the creation of explicit technology road maps for at least some of the programs of the Office of Nonproliferation Research and Engineering will most likely be required, at least in the near future. Fortunately, most of the information required to develop these road maps appears to already be available to program managers within the Office. As with many of the road maps in the Department of Energy Strategic Plan, the missing piece appears to be the definition of important thresholds for performance. Performance thresholds should enable a solution to a specific operational need. Quantifying these thresholds is the most important output from a Concept Option Group. If cross-functional teams are convened using some other planning methodology, particular care should be taken to demand that the teams specify such thresholds. The addition of a process to quantify those thresholds would prove valuable in producing truly useful technology road maps.

Additionally, a technology road map including such thresholds would be very useful to a program manager directly involved in a technical area. It would provide the manager with detailed milestones for monitoring the progress of research activities. Incipient problems would become obvious sooner rather than later and corrections to fix the problems could begin earlier, lessening their effect.

The inevitable trade-off is the time that researchers and managers must spend in crafting the road map. But as long as such road maps are demanded by higher levels of management, the Office of Nonproliferation Research and Engineering should make them useful by including important technical thresholds identified through a process such as the one I described earlier. In particular, road maps seem to be most useful when particular initiatives demand the coordinated accomplishment of multiple technical objectives. I expect this will most likely be the case in large-scale developments, such as system development. In that case, the Office might choose to create a road map for coordinating its own activities, independent of any external demand.

Planning Methods Developed for Business

The managerial literature is filled with planning methods developed for business. Indeed, it seems that every management consultant with a new technique publishes a book hoping to prompt a new fad in strategic planning. Although valuable in a certain sense, these techniques all share the same problem within the context of a governmental research program. As one author noted, "The goals inevitably are reduced principally to financial terms . . ." (Steele, 1989, p. 204). This approach is natural for a profit-making organization. The bottom line is real and very quantifiable. Although the managerial literature frequently notes the difficulty of connecting various business activities to profits, most business activities *can* be linked to producing a profit. The research activities of businesses, even in the later stages of development that lead to actual production, can be connected in a straightforward fashion.

This business context is quite different from that of most federal research programs. Most federal research is done to pursue societal goals that are not economic in nature. Investments in health research, for example, are made to improve our quality of life, and provide an economic benefit only indirectly. Investments in space exploration or high-energy physics are justified by their bettering our understanding of the universe, and not for their economic benefits. In the case of the research in the Office of Nonproliferation Research and Engineering, its purpose is to enhance some aspect of national security, with the linkage to specific national security objectives explained through an objective-based planning method, discussed earlier in this chapter. Consequently, the elaborate methods that attempt to connect research expenditures to sales, profits, or even stock prices for businesses (Industrial Research Institute Workshop, 1994) are not applicable to most governmental research programs, including that of the Office.

In fact, business methods that attempt to connect research expenditures to profits cannot be applied to the Office's research program even through an analogy because there is no single metric that operates as a "bottom line." National security encompasses many objectives and responds to many threats. The importance of any research program depends in part on the importance of the objectives it supports, which is inherently a subjective judgment. Because research work in a public agency, such as the Office of Nonproliferation Research and Engineering, lacks the single metric that economic benefit provides to a business, methods used in business to link dollars spent on research to profitable returns can offer only insights, and not an analytic tool, for this research program.

Balancing the Research Within a Program

The lack of a single metric that can characterize the output of a research program has another important implication—it makes it difficult to balance the research program as a "portfolio" across its various dimensions (I discuss this concept further in the next section). Without a single figure of merit or metric to measure the "goodness" of a choice, there is no way to analytically balance the program. Rather, balancing the Office's research and development program across missions, users, and time horizons critically depends on the subjective judgment of the Office leadership.

In striking this balance, the leadership has to consider many factors beyond those that are intrinsic to the research itself. As noted earlier, the Office's research program is the steward of the NN Tech Base and, in part, of the national laboratories within NNSA. Some of its research can and should be justified simply to fulfill this stewardship role. Additionally, changes in the research program must be incremental as research institutions legitimately take time to adjust to significant changes in research direction. If institutions are subjected to frequent changes in direction, the result can be wasted effort and poor morale. Frequently, the choices end up focusing on small changes in the existing plan, such as what should be cut from or added to an existing effort. Planning methods should be designed to inform such choices by making program managers aware of not just the technical quality of the research efforts but their importance to certain users and therefore to a certain national objective. Finally, and obviously, the program must implement the choices made at the NNSA or national level. It is at those higher organizational levels that priorities among, or the balancing of resources across, very different objectives are set.

Lessons from the Business Literature

I have already drawn on two insights from the business literature without identifying them as such. Both the idea of a "portfolio" of research and the concept of balancing such a portfolio across measures of risk and a time horizon come from the business literature (Steele, 1989, pp. 192–194). Often, a portfolio is coupled with a simple matrix in order to identify the most promising or profitable research programs. Typically, the axes of the matrix are labeled with terms such as "business position" and "industry attractiveness" (Steele, 1989).

Displaying the elements of a research program through such a matrix can illustrate the scope and range of a diverse program. Economically motivated dimensions for a matrix are of no use to the Office's research program, but alternatives can be fashioned. RAND, for example, has used noneconomic dimensions in similar matrices to illustrate the scope and range of military research programs. Both the terms "importance to the eventual customer" and "generality of application" were used in analyses for the U.S. Navy and U.S. Army (Saunders, et al., 1995, and Wong, et al., 1999).

Matrices can be useful in two ways. First, they can aid the managers of the research program in explaining to diverse audiences the scope of the Office's efforts. Second, the matrices can help in choosing the most promising research programs, for example, by helping to identify initiatives that are of the greatest importance. In both cases, the utility of the matrix depends upon believing the estimates of "importance" or whatever terms are chosen to characterize the matrix axes. Once again, the positioning of a research effort on the axes relies upon the subjective judgments of informed users and developers.

As in the earlier discussion of technology road maps, these matrices, or diagrams, can be constructed at various levels of aggregation. At one extreme, each project can be subjectively placed within the matrix; at the other extreme, entire programs can be placed only roughly to indicate their predominant efforts. In choosing the level of aggregation, the real cost in time and effort of fashioning such matrices should always be weighed against the expected value. Understanding and evaluating each project, even subjectively, is typically a time-consuming task. In practice, such matrices may be of most use to a new program or a program undergoing dramatic shifts for one reason or another. In those cases, each potential project might be subjectively placed within the matrix. The managers of the program would then be able to pick among the potential projects, fashioning a program that is balanced across the overall dimensions of the matrix.

Another insight from the economic literature on business-related research and development seems quite applicable to all research. Analyses of the financial worth of patents and other innovations have indicated that most financial gains come from a small percentage of all patents, less than 10 percent of the total (Scherer and Harhoff, 2000). This fits with my personal observations of federal national security research. Even highly successful programs, such as those of the Defense Advanced Research Projects Agency, have more failures (e.g., unsuccessful efforts to develop useful battlefield robots or electromagnetic guns) than successes (e.g., low-observable aircraft). If federal programs match the characteristics of the research sample of Scherer and Harhoff, the difficulty in maintaining support for publicly funded programs is in explaining why 90 percent of a program's efforts fail to lead to useful systems. This would be especially true if simple statistical fluctuations cause the results to be notably worse than average for some years, as they did for Scherer and Harhoff. Referring to these findings may help explain such apparent failings.

Other insights from the business literature include some general lessons learned about the application of planning methods. For example, the literature emphasizes that *any* method should avoid seemingly endless demands for data or people's time (Steele, 1989). In a sense, this elaborates a point made earlier—there is an opportunity cost in the time and effort expended in any planning activity. From the very beginning, care should be taken to avoid allowing a planning activity to grow without any limitations on it. Another lesson from the same author is that strategic planning of any sort must be used by management in a meaningful manner (Steele, 1989). Simply creating a strategic plan, road map, or matrix has little worth in and of itself. Only when managers actually use the results of their planning tools will those results be worth all the time and effort put into achieving them.

A final lesson from the business community involves users. When research and development is conducted within a business firm, identification of the actual customers for the research is critical to directing the research, and it is essential that the research and development are guided by the *users'* values, and not those of the researchers. This advice grew out of an energy research project at a large oil company. In particular, gaps in capabilities perceived by internal company customers were much more effective at spurring eventual use of new technologies or systems than were the technical opportunities seen by company researchers (Hirsch, 2000). This insight appears to be directly applicable to the Office's research program because it addresses the willingness of a user to adopt a new technology or embrace a new concept, which is independent of whether the benefit to the overall institution is an economic one. Instead, the user's acceptance or rejection of new technologies or concepts is based on the common human reaction to new ideas. In part, this reaction is what prompted the creation of Concept Option Groups, which involve users. The reaction is quite universal, though, and must be considered whether or not a Concept Option Group is involved.

The Government Performance and Results Act and Related Mandates

Independent of the mandates that the NNSA or the Department of Energy places on the particular planning methods of the Office of Nonproliferation Research and Engineering, the Office must also satisfy congressional requirements commonly placed on all federal programs. The most recent large changes in those demands are the plans and reports mandated by the Government Performance and Results Act (GPRA) of 1993. This act mandates the creation of strategic plans, annual performance plans, and annual performance analyses that are tied to program outcomes for almost all government programs (National Science Board, 2000, pp. 2–14). Congress recognized that research and development programs might present particular problems in defining quantifiable metrics for their outcomes, which has proved to be true in practice.

In response, various panels of experts have provided advice on how the GPRA should be applied to research programs. Although the annual, formal response to GPRA will be coordinated at a high organizational level within the NNSA or the Department of Energy, the Office of Nonproliferation Research and Engineering will certainly support the formal response and therefore also should be familiar with that expert advice.

The most widely publicized advice on adhering to GPRA within a research program has come from the National Academy of Sciences. Much of this advice concentrates on the "basic" research effort, for which the conceptual problems are most difficult to solve, but the Academy report does cover applied research as well (Committee on Science, Engineering, and Public Policy, 1999). Most of the programs within the Office can therefore benefit from the National Academy's advice.

The most important part of the National Academy's advice relates to the use of peer review of the quality of the research and relevance assessments. The National Academy panel recommended these reviews not only for basic research areas but also for large demonstrations and applied research. The terms the National Academy uses for these reviews are "general relevance," "high quality," and "best people" (Committee on Science, Engineering, and Public Policy,

1999). In any case, these reviews are admittedly subjective judgments, made by an assembly of experts. To avoid bias, it is essential that the panels be composed of individuals who are outside of the program. It would seem to benefit the Office if it implemented a greater number of these reviews.

On applied research programs more specifically, the National Academy report largely assumes that the annual GPRA reports from each agency could readily draw upon existing milestones and similar existing outputs within each research program, using such outputs as definable metrics. Such milestones may exist for some of the Office's larger demonstrations, such as the satellite programs, and so should be used to help satisfy the GPRA reporting requirements, but may not exist for some of the smaller applied efforts that are not focused on a demonstration. In such cases, the Office must then fall back on the peer reviews.

A final point drawn from the National Academy Report is the importance of human resources—the scientists and engineers who perform the research. The report is primarily concerned with the production of graduates from research universities. The Office of Nonproliferation Research and Engineering has only a small role to play within the research universities. However, the Office plays a large role in another aspect of human resources—the unusual, if not unique, training it supports in the NNSA's weapons laboratories. Because of the security that is required, this training can be conducted only within those laboratories and, as such, is the sole responsibility of NNSA. This training is also a part of the NN Tech Base mentioned earlier in this chapter (U.S. Department of Energy Nonproliferation and National Security Advisory Committee, 2000, p. 25f). While not explicitly part of the GPRA reporting requirements, including some estimates of the new technical manpower the Office has created through its activities may serve its interests well. Inclusion of these estimates in the GPRA report would be responsive to the interests of outside groups, such as the authors of the National Academy Report or the Chiles Commission (Chiles, et al., 1999).[2] It would also help to ensure that the Office's research program will have the technical expertise it needs in the future.

Established processes exist for expert panel reviews of the relevance of a research program to the overall goals of an organization. Perhaps the most documented processes are those of the U.S. Navy, in which panels are explicitly organized

---

[2]The Chiles Commission on Maintaining United States Nuclear Weapons Expertise recommends specific strategies for recruiting and retaining the scientific, engineering, and technical personnel needed to maintain a safe and reliable nuclear weapons stockpile without engaging in underground nuclear testing. Chaired by Admiral H. G. Chiles, Jr. (U.S. Navy, retired), the Commission was established by Congress under the National Defense Authorization Acts of 1997 and 1998 (see http://energy.gov/HQPress/releases99/marpr/pr99037.htm).

and tasked to estimate the relevance to the Navy of the research programs that the Navy funds (Kostoff, 1988). In their form, these panels resemble the expert peer review panels assembled by the National Science Foundation or the National Institutes of Health, with one difference. They include active-duty officers, who are not part of the research community, to formally estimate relevance. Presumably, the National Academy panel had a organization in mind for the relevance review it recommended, but the panel did not include any details of its implementation in the Committee on Science, Engineering, and Public Policy (1999) report.

3. Suggested Enhancements to Program Planning

Two salient themes appear consistently across this review of strategic planning methods:

- Maintaining links to the end users of the research program
- Independently reviewing the quality of the technical research supported by the Office.

Both of these themes are important for two reasons: (1) They would improve the actual systems and technologies developed by the Office of Nonproliferation Research and Engineering, and (2) they would increase external support for the program through involving potential users who understand the ultimate applications for the research. Support from outside users is the only answer that effectively resolves questions about the utility of a research program. Without such support, any program can come to be regarded as self-serving and have its funding reduced, whatever its true worth.

Applying these two themes to the Office's research planning might also prompt changes in the content of the research by identifying new opportunities of interest to potential users. Happily, both themes can be encompassed within a common framework.

Connecting to Diverse Customers

By using an objective-based planning structure to stay connected with its diverse customer base, the Office could more effectively link all of its projects to national or departmental objectives. Such a structure would naturally identify specific (or representative) customers for research products. Creating such a planning structure also helps to identify many of the holes (e.g., apparently unaddressed needs) and overlapping areas (e.g., two separate R&D programs supporting very similar work) that others might note. Often, an apparent hole is filled by an existing research effort within another organization that is known to the researchers in the technical field, or an overlap turns out to be more apparent than real. It is important to understand such issues before others raise them and then having to react to them. However, the precise method that is used to link the

Office's projects to national goals is less important than that some consistent organizational structure be used.

The next step, after creating an objective-based planning structure, is to involve the end users in the planning of the program, whether they are identified through the objective-based planning method or are already known to the program managers. In practice, users will spend a significant amount of time with the research program's managers only if the effort results in changes in the program that are important to them. One effective way of identifying such changes is though a COG-type panel in which the users help to identify new initiatives for a program. Another, perhaps less exciting, role for users is to be part of an expert panel providing relevance review. The COG-type panel is better suited to programs undergoing change whereas the expert panel can be made part of the regular reviews of a continuing research program.

With either type of panel, it is essential to take the opinions of users seriously when setting priorities. Nothing dampens a user group's enthusiasm more than the perception that the panels and reviews are only for show, and meanwhile the real decisions are being made elsewhere. The willingness to allow outside users to visibly influence internal decisionmaking is often a difficult step for a research organization to take, but it pays off by greatly increasing the utility of the research.

Some of the planning methods discussed in Chapter 2 can be used within the objective-based planning process or COG process. For example, a matrix produced for some other purpose can be also be used as a visual aid in relevance reviews or even as part of a methodology for judging the relevance of the research to the user community. Likewise, a technology roadmap can be used with a COG to illustrate both the difficulty inherent in, and the potential benefits of a technology research plan. These methods can enable the combined group of users and technical experts to better choose among competing concepts.

Maintaining and Demonstrating Quality

In addition to establishing and maintaining connections with users, I believe the research program of the Office of Nonproliferation Research and Engineering must add enhanced peer reviews to its research planning process. Both the quality of the planned research and the quality of the researchers themselves should be subjected to peer review. Such reviews enhance the argument that the program supports demonstrably high-quality work, which could also help in the budget debates within Congress, the Office of Management and Budget, and the Department of Energy. Besides incorporating the recommendation of the

National Academy of Science to include such reviews, (Committee on Science, Engineering and Public Policy, 1999), such reviews are also in keeping with established guidance from the Clinton Administration (Executive Office of the President, 2000). While that guidance may eventually be modified by the current White House administration, the importance of peer reviews for assuring high-quality research is unlikely to be diminished. That aspect of science policy seems to be bipartisan and noncontroversial.

To be effective, the peer reviews must be visible to those outside the Office and involve outside experts. For example, the Chemical and Biological Defense Program within the Office has already made use of peer review panels, and its experience with those panels and methods of peer review could provide a template for other programs within the Office. The majority of the members of any review panel must *always* be from outside the national laboratories. Experts from outside the Department of Energy complex, for example, should be sought to avoid any appearance of bias toward the national laboratories.

Peer review panels can easily be expanded to include relevance screening performed by selected user representatives. User panels frequently benefit from involvement in merit reviews by gaining a deeper appreciation of the substance of a proposal. Likewise, insights from users can help merit reviewers determine whether the research is targeted correctly. Finally, combining these two types of reviews (peer and user) into a single, integrated process should lessen the total burden on the researchers who must explain their work.

4. Advantages of the Recommended Enhancements

The two enhancements to the Office's research program planning that are recommended in this report—establishing connections with the customers of the research products and conducting independent reviews of the quality of the Office's research—have five principal advantages.

1. Consistency with Existing Methods

The two proposed enhancements are consistent with the existing research planning methods used in the Department of Energy. National objectives are already cited in the Department of Energy's existing strategic plan, which naturally lead to involving the appropriate users. More-detailed aspects of the existing planning methods are also consistent with the two proposed program-planning additions. For example, the concept of "R&D portfolios" is a key principle within the Department of Energy that the Office of Nonproliferation Research and Engineering can also use and even expand upon. The Office can justifiably claim a "portfolio of portfolios" across customers, technical areas, and types of research, as discussed earlier in Chapter 2. Additionally, tools such as technology roadmaps that are used widely for planning R&D, even outside the Department of Energy, can be used to support the proposed enhancements.

The quality review also echoes a departmental theme: the emphasis on "science" as the core of the Department of Energy and presumably of the NNSA as well. Consequently, there should be no objection to reviews aimed at enhancing quality nor to those reviews making that improved quality more visible to audiences outside the department.

2. Increased Visibility of NNSA Customers

The proposed approach can help the Office to better describe its role to the NNSA, which in turn would help to identify customers the Office serves who may be unfamiliar to the rest of the NNSA. With the majority of NNSA's funding, and perhaps the majority of its attention, focused on the demands of its stewardship of the national stockpile of nuclear weapons, explicitly pointing out these users could be important in drawing enough attention to these valid needs.

Additionally, the creation of visible links to users should influence any planning, programming, and budgeting–type system created by the NNSA, ensuring that users' needs are not overlooked. It is important that any such system explicitly include the needs of these users. Otherwise, the dynamics of the planning system would discourage funding of research for users not within the system and favor those the system does include.

3. Matched Trends Across Government

The proposed approach reflects current trends in managing research and development across the federal government. This is true for both GPRA-related reporting and also general guidance on research from the Office of Management and Budget. Moreover, by following the suggestions of the National Academy of Science panel, the approach would have a strong intellectual basis.

4. Improved Utility of Research Results

The proposed approach offers the potential to increase the actual utility of the research, which is the aim of the program. Establishing a connection to users of the products of the research should avoid an all-too-common phenomenon—the rejection of new alternatives simply because they are unfamiliar. Involving users early in the development of a concept helps to avoid the "not invented here" attitude that leads to rejection of an idea simply because someone else proposed it. Ideally, user involvement may even generate more and better ideas.

5. Alternatives Provided for Program Management

Finally, the peer review, and perhaps also the relevance review, could provide a set of unfunded efforts that can later be tapped by individual program managers when additional funding becomes available. The reviews would also identify the lowest-rated funded projects for the individual program managers, which is useful information if a budget cut must be absorbed by a program. Consistent peer review should result in more-productive discussions among program managers and Office leadership on allocating cuts or additional funding across programs. All this information should make the management of adjustments in project funding an easier task.

The allocation of budget adjustments stems from the subjective judgment of the leadership of the Office of Nonproliferation Research and Engineering. There is no magic scale or metric that indicates whether more money should go toward designing smaller detectors for chemical agents, for example, or toward developing cheaper nuclear detectors. Real uncertainties about the future will almost always leave such choices equally defensible. The proposed program changes suggested in this report would, however, give Office management a better understanding of both the contribution each research effort makes toward some operational end and the technical quality of the work. This should, in turn, enable more-informed choices to be made initially when compiling a budget, and better choices to be made incrementally when making quick budgetary adjustments.

Bibliography

Beattie, C. J., and R. D. Reader, *Quantitative Management in R&D*, London: Chapman & Hall, Ltd., 1971.

Camp, R. C., *Benchmarking—The Search for Industry Best Practices That Lead to Superior Performance*, Milwaukee, Wis., ASQC Quality Press, 1989.

Chiles, H., et al., *Report of the Commission on Maintaining United States Nuclear Weapons Expertise*, Washington, D.C.: General Services Administration, 1999.

Committee on Science, Engineering, and Public Policy, *Evaluating Federal Research Programs*, Washington, D.C.: National Academy Press, 1999.

Davis, P. K., and Z. M. Khalilzad, *A Composite Approach to Air Force Planning*, Santa Monica, Calif.: RAND, MR-787-AF, 1996.

Dewar, J., C. Builder, M. Hix, and M. Levin, *Assumption-Based Planning: A Planning Tool for Very Uncertain Times*, Santa Monica, Calif.: RAND, MR-114-A, 1993.

Executive Office of the President of the United States, *Government-Wide Performance Plan, Budget of the United States Government, Fiscal Year 2000*, Washington, D.C., 2000.

Hirsch, R. L., *Capturing the Benefits of Applied Research & Development: Insights from Experience & Case Studies*, Washington, D.C.: Advanced Power Technologies, Inc., January 2000.

Industrial Research Institute Workshop, *Technology Value Pyramid Menu of Metrics and Definitions for the Menu*, Washington, D.C.: Industrial Research Institute, 1994.

Kent, G. A., *A Framework for Defense Planning*, Santa Monica, Calif.: RAND, R-3721-AF/OSD, 1989.

Kent, G. A., and W. E. Simons, *A Framework for Enhancing Operational Capabilities*, Santa Monica, Calif.: RAND, R-4043-AF, 1991.

_____, "Objectives-Based Planning," in Paul K. Davis, *New Challenges for Defense Planning—Rethinking How Much Is Enough*, Santa Monica, Calif.: RAND, MR-400-RC, 1994.

Kostoff, R. N., "Evaluation of Proposed and Existing Accelerated Research Programs by the Office of Naval Research," *IEEE Transactions on Engineering Management*, Vol. 35, No. 4, 1988, pp. 271–279.

Mintzberg, H., *The Rise and Fall of Strategic Planning*, New York: The Free Press, 1994.

National Science Board, *Science and Engineering Indicators 2000,* Vol. I, National Science Foundation, Arlington, Va.: U.S. Government Printing Office, 2000.

Office of Deputy Administrator for Defense Nuclear Nonproliferation, "Multispectral Thermal Imager Satellite," http://www.nnsa.doe.gov/about_nn.asp (last accessed October 26, 2001).

R&D Productivity—Study Report, Culver City, Calif.: Hughes Aircraft Company, 1974.

Saunders, K. V., et al., *Priority-Setting and Strategic Sourcing in the Naval Research, Development, and Technology Infrastructure*, Santa Monica, Calif.: RAND, MR-588-Navy/OSD, 1995.

Scherer, F. M., and D. Harhoff, "Technology Policy for a World of Skew-Distributed Outcomes," *Research Policy*, Vol. 29, Nos. 4–5, 2000, pp. 559–566.

Schwartz, P., *The Art of the Long View*, New York: Doubleday, 1991.

Skiby, J., ed., *Nonproliferation and International Security 1998*, Los Alamos, N.M.: Los Alamos National Laboratory, February 1998.

Steele, L. W., *Managing Technology*, New York: McGraw-Hill, 1989.

Thaler, D. E., *Strategies to Tasks: A Framework for Linking Means and Ends*, Santa Monica, Calif.: RAND, MR-300-AF, 1993.

U.S. Department of Energy, *Implementation Plan National Nuclear Security Administration,* Washington, D.C., 2000a.

_____, *DOE Research and Development Portfolio: Overview*, Washington, D.C., 2000b.

_____, *DOE Research and Development Portfolio: National Security*, Washington, D.C., 2000c.

_____, *Department of Energy Strategic Plan*, Washington, D.C., 2000d.

U.S. Department of Energy Nonproliferation and National Security Advisory Committee, *DOE Research and Technology Against the Threat of Weapons of Mass Destruction,* Washington, D.C., 2000.

Wack, P., "Scenarios: Uncharted Waters Ahead," *Harvard Business Review*, September/October 1985a.

_____, "Scenarios: Shooting the Rapids," *Harvard Business Review*, November/December 1985b.

Wong, C., et al., *An Approach for Efficiently Managing DOD Research and Development Portfolios*, Santa Monica, Calif.: RAND, RP-791, 1999.